Simplifying Cloud Computing for Everyone

Demystifying Basics

Pankaj Gupta

Copyright @Pankaj Gupta, Dec.2022

Dedication

This book is dedicated to people who want to understand the concept of cloud computing services and business model. It simplifies the complexity of cloud for everyone who just want to understand the basics of the cloud computing to enhance their knowledge. This can also be used for training purpose for business people to give them a conceptual understanding of cloud services.

Cloud Contents

Chapter-1 6

The Concept 6

What is Cloud? 6

Cloud Business Model 9

Internet 10

Explaining Cloud Computing 13

Advantages of Cloud services 18

Chapter-2 21

Cloud Computing Services 21

Infrastructure as a Service (IaaS) 22

Platform as a Service (PaaS) 23

Software as a Service (SaaS) 24

Chapter-3 27

Cloud Deployment Models 27

Private Cloud (On-premises) 29

Community Cloud 31

Public Cloud 31

Chapter-4 34

Related Terms 34

Data Centres 34

Virtualization 36

Internet of Things (IoT) 38

Official Definition of Cloud Computing 39

From the Author 43

Related books 44

Chapter-1

The Concept

What is Cloud?

You all have heard of clouds. Cloud is formed of water droplets; the cloud is aggregation of water droplets . So, when cloud rains it drenches everything in the area below it. That means it's a common resource which is available to people in shared environment, although it's a natural resource. Now the same concept is replicated in cloud computing business model where shared resources are made available to everyone. You just need to access it.

Let me simplify it in terms of real-life utilities. We all have been using Electricity services at our home and offices for decades. How do we get electricity connection?

To subscribe, initially we pay one-time installation charge for the minimum hardware required at our premises i.e., wiring and meter. Then we have to be wired to the nearest transformer installed on an electrical pole which is further connected to main powerhouse. Thereafter we only pay for the minimum subscription charge (monthly rental) and the usage charges.

Similar is true for water supply services also. In olden days we used to dig well in our homes or at a common place in the village. Then we had to fetch water from the water well and stored in buckets. With development in civil engineering, we constructed water tanks in our colonies and created a common storage of water which became a shared resource for everyone in that area.

Here also we just need to install minimum hardware required at our premises and get connected to the nearest water tank through water pipes then we start getting continuous water supply. Further, we only have to pay a nominal amount as monthly rental and bill as per usage.

Cloud service model is analogous to these traditional electricity and water services where we minimise the hardware required at customer premises and centralize the utility resources at a common place usually called "Data Centre". This leads to migration from "CAPEX business model" to "OPEX model" i.e., reducing the infrastructure cost (Capital Expenditure-CAPEX) for users and making shared resources available at nominal recurring cost (Operational Expenditure-OPEX).

Cloud Business Model

You all must be using various applications like emails, office outlook 365, CRM etc., on daily basis. Well, the concept is similar. You have a hardware device at your home or office (Mobile phone, Laptop or Desktop Computer) and a wireline or wireless connection for accessing the servers installed at any service provider's premises i.e., data centres in cloud environment.

This way you get access to storage space, files, application softwares and databases as per your need. Thus, it is on-demand and pay per use service on contrary to traditional investment in IT infrastructure and fixed operation cost.

Internet

We all have been using web browsing and web-based email services through Internet for more than a decade now e.g., outlook email, Gmail, yahoo email, are a few but there may be many more. But what is Internet actually?

Internet is a network of networks. Now what is a network?

Network means added or connected devices like computers, servers, or any other element. Internet started with connection of four computers in a lab for a single organization but expanded further into a global network i.e., it is a collection of computers spread across globe but interconnected with any of the transport mediums (Copper, Fibre, or Satellite).

To make it easy to understand, let me put this way "Computer network is similar to our people's network connected through any social media platform. And networking means adding computers like we add people in our social network."

Suppose you have a family of four person in a house connected with each other and further you are connected with three other families of four persons. So, this network of families can keep expanding. In the same way Internet has been evolved and it has become network of networks connecting millions of elements now.

Internet is also represented as cloud since it is formed of millions of computers and shared by people globally. In a simple way, it is depicted in the figure below.

Figure-1: A representation of Internet

Explaining Cloud Computing

Let me take the example of email services again. We have been using the webmail services (e.g., yahoo mail, Gmail etc.,) for personal use as well as for small businesses. These web email services are very good example of cloud services where we just need to connect to email server on internet, then login using our user's name and password and then we can use the email services; reading mails, sending, and receiving mails etc.

In past, enterprises had to set up their own mail exchange server and maintain it and have their own control. Traditionally enterprises had been investing in creating IT infrastructure at their premises and maintained it. They usually identified a separate room to install IT equipment and they called it Server room or Computer Room or Communication room as per prevalent practice, available technology, and privacy. For all these practices an enterprise had to invest in their own infrastructure and hire the IT resources to maintain and manage it, so this is CAPEX model.

For example, In late 1990s I worked in an organization where we created our own infrastructure of email server, domain name server and web server. So, we spent a huge amount in purchasing, installation, and commissioning of these elements. This was an example of Capital Expenditure.

Development of Data Centres provided an option of Co-location services where enterprise could rent power and space in the service providers premises and created their own network infrastructure off-site in data centres and accessed remotely through internet.

"Now with the evolution of cloud concept, service providers have created a common infrastructure of email servers, domain name servers, web servers, storage and databases where they invested in these elements and any organization needed theses services can subscribe and share the same infrastructure."

So, now enterprises and users only need to subscribe to ready-to-use infrastructure and pay only for the subscription and usage charges as per their usage hours or whatever is the billing unit. Thus, users have to bear only operational expenses. This is OPEX model.

Thus, **Cloud Computing offers on demand services for computing, storage, software, and other applications with usage-based billing.** Using services in cloud is essentially migration from CAPEX to OPEX model.

Now, what does a user need to reach cloud?

An access medium (like wireless, copper or fibre cables) to reach the shared infrastructure in cloud. Then using their mobile phone, laptop, or desktop and through internet they can reach the IT infrastructure in the cloud.

A simple demonstration of the cloud services model is shown in the figure below.

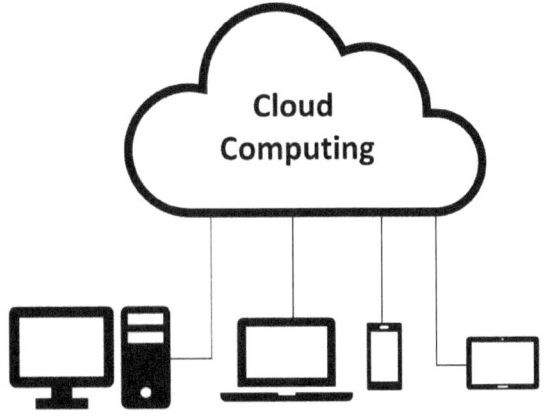

Figure-2: Cloud Computing Concept

Wish we have an access medium to reach the clouds in the sky so that we could make them to rain whenever we needed. ☺

Advantages of Cloud services

1. Small businesses don't have to purchase so many computers and servers to make their IT-infrastructure as they can subscribe to already existing infrastructure in cloud and pay as per usage.
2. Individual users also don't have to buy hard disks to store their data, they can subscribe to any service provider and use storage space from their shared infrastructure in cloud. Best example is Google Drive storage space which we get with Gmail.
3. Enterprises can subscribe to emails and web hosting without

creating their inhouse email server and web server.
4. Only minimal hardware and software required at customer's premises i.e., laptop, desktop, tablets or mobile phones so very low expenses and they only have to pay for a nominal subscription charge and/or usage charge.
5. Upgrades and downgrades are easy as you just have to place order with service providers instead of buying additional hardware and software in CAPEX model.
6. Service commissioning is faster as all the resources are already in place with service providers. They only need to add you as

user and grant access to the required resources.

7. Users don't have to employ IT-support staff to maintain their in-house infrastructure like CAPEX model. A common resource of IT-staff at Data Centre maintain the IT-Infrastructure and they can also provide remote support for users in need.

Chapter-2

Cloud Computing Services

What are the types of cloud computing services or business models?

Well, there are different ways we can use the cloud computing services.

- Infrastructure as a Service (IaaS)
- Platform as a Service (PaaS)
- Software as a Service (SaaS)

Let's know a little bit about these options.

Infrastructure as a Service (IaaS)

It can provide you whatever computing infrastructure you need in cloud environment like physical hardware with operating systems, virtual servers, storage space, and database connections. It can also provide you separate dedicated hardware, software, and storage as per your need.

This service gives you flexibility and management control over your IT resources although physically hosted in service provider's cloud. You don't have to purchase the resources and bother for installation and maintenance.

For example, Amazon WorkSpaces provides virtual desktops in the cloud for End user computing.

Google Drive is another example of IaaS where you can store your files in the cloud and share with others.

Platform as a Service (PaaS)

It provides tools and platform for software developers and application programmers in cloud environment for creation of application, webpages, and other softwares. For example, if you want to develop your own website then you can subscribe WordPress services to use their website builder and design tools for website development.

This service helps you to focus on management of your applications only as you don't have to worry for procurement of hardware, software and updating patches. For example, Amazon

Developer Tools provides software development service on AWS.

Software as a Service (SaaS)

Any software application you access on Internet and start using is like SaaS. Web-based emails, MS-Office 365, Salesforce are few examples. So, instead of buying the software you only subscribe and pay for use.

These are ready to use applications or software from user perspective. Users do not have to download and install email client on their device. They do not have to maintain the software application. They just have to use internet browser to access the email service.

For example, I am using Gmail service which I can open anywhere on any device; mobile phone, laptop, or desktop. I just need internet connectivity to access the Gmail server. Google Docs is another example where you can create word document and share with others for their inputs and modification.

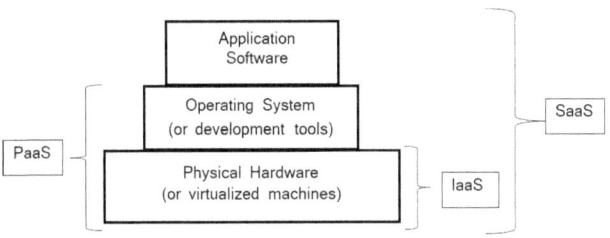

Figure-3: Explaining Cloud Services

Figure above shows a simple representation of the cloud service models. In IaaS, you only use infrastructure, in PaaS you use infrastructure and development tools and in SaaS you use application software in addition to previous two layers. But as a user you don't have to worry about the underlying hardware and operating systems.

Chapter-3

Cloud Deployment Models

Various cloud service delivery models only differ by their deployment strategies; how the cloud infrastructure is deployed, though the concept of the cloud will remain same.

Let's understand in a simple way. Different cloud models are defined by physical boundaries like you have a small housing society made of few hundred houses and there is a pool of common facilities like club, sports ground, play school etc., are available to local residents. Then you have a bigger society made of thousands of houses with pool of common resources available to share by society people. And then you have a city made of so many societies and colonies with pool of common resources available to all public on subscription basis.

It is somewhat similar to network topologies like Local Area Network (LAN) which belongs to a single organization, Metropolitan Area Network (MAN) which connects a city and Wide Area Network (WAN) which spreads globally. Likewise, we have Private Cloud, Community Cloud and Public

Cloud. Let's understand a bit about them.

Private Cloud (On-premises)

In this model, IT infrastructure is made for a single organization and used for inter-office business transaction's purpose. Here the computing resources are owned and operated by a single organization and not shared by other customers. It can be at the customer premises or in the data centre. So, it's almost similar to the traditional IT infrastructure but with the benefits of cloud computing features.

Figure below depicts the concept of private cloud.

Company A   Company A's Private Cloud

Figure-4: Private Cloud

Private cloud is preferred by the companies who need confidentiality and deal with sensitive data or to meet their regulatory compliance. It may be compared to your private locker in a bank where you only have the key and access the same though its physically located in bank.

Community Cloud

In this model, the computing resources are made available to a specific community or a group of organizations. A shared IT-infrastructure is set up in data centres in cloud environment for a group of companies with similar interests like a chain of hospitals or offices of a banks or financial institutions or any other business houses.

Public Cloud

Here the IT-infrastructure is made available to all public and deployed by government, educational or a business organization. Cloud Service Providers (CSP) own and maintain the infrastructure and access to the resources is offered to multiple customers on subscription basis or pay per use billing. Cloud provider handles all the privacy, and security issues for all subscribers.

Figure below shows a public cloud shared by many companies.

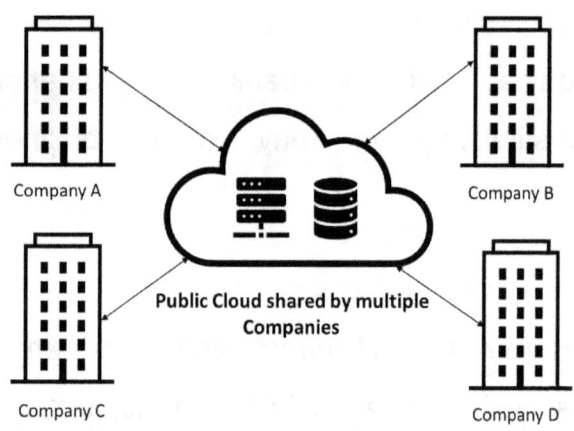

Figure-5: Public Cloud Model

Just to share that there is also a fourth option of "Hybrid Cloud" which is a combination of two or more cloud model which I am not discussing here to maintain simplicity.

Chapter-4

Related Terms

Data Centres

Data Centres are essentially part of cloud; a cloud may be formed by one or more data centres connected through any transport medium like national and long-distance fibre links, International submarine cables or satellite links.

Data centres may have physical computers, servers, storage devices like hard-disk system with security mechanism to protect from unauthorised access virtually as well as physically, i.e., data centre facilities are in highly protected area with multiple level of security checks in place while entering the building premises to allow physical access to authorized persons only.

A partial view of data centre facility is shown below.

Figure-6: Data Centre

Virtualization

With development of virtualization technology, it is now possible to create multiple Virtual Machines (VMs) and virtual server on a single physical machine hardware. Each virtual machine works like an independent computer system or server with its own operating system and application softwares though hosted on a single physical system.

Required resources are allocated to virtual machines from the physical system through a software called hypervisor.

Figure below shows a simple representation of virtualization concept where multiple Virtual Machines (VMs) are created on a single physical system.

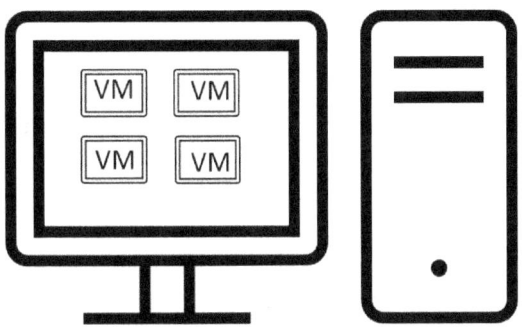

Figure-7: Virtualization Concept

Virtualization reduces the need of physical hardware and it is quick to create the resources and upgrade/downgrade as per customer needs. It also reduces the cost of IT infrastructure for service providers as well as running cost for customers.

Internet of Things (IoT)

IoT term refers to the aggregated network of collective devices spread across locations (home, business, or offices etc.,) which communicate with the help of technology and facilitate remote monitoring and management. Things to be monitored are embedded with sensors and software for connecting through internet to communicate their status. All the end devices are connected to Wi-Fi or a secured Local Area Network (LAN).

IoT extends the internet connectivity to variety of devices at home and offices which can be monitored and controlled remotely thus reducing the human efforts and cost of in person monitoring.

Figure below shows a representation of IoT model.

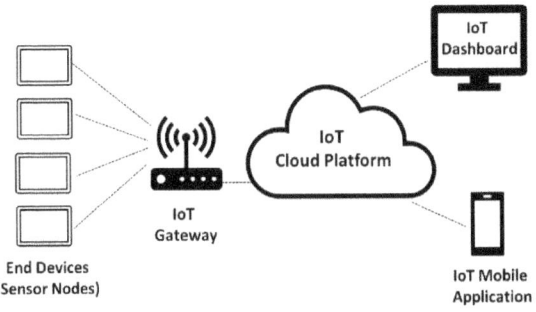

Figure-8: IOT Model

Official Definition of Cloud Computing

Let's become a bit technical for a moment. As per National Institute of Standards and Technology's (NIST), here is an official definition of Cloud Computing:

"Cloud computing is a model for enabling ubiquitous, convenient, on-demand network access to a shared pool of configurable computing resources (e.g., networks, servers, storage, applications and services) that can be rapidly provisioned and released with minimal management effort or service provider interaction."

The NIST definition lists five essential characteristics of cloud computing: on-demand self-service, broad network access, resource pooling, rapid elasticity or expansion, and measured service. It also lists three "service models" (software, platform, and infrastructure), and four "deployment models" (private, community, public and hybrid) that together categorize ways to deliver cloud services.

Reference: https://www.nist.gov/news-events/news/2011/10/final-version-nist-cloud-computing-definition-published

Future of cloud computing

In the past decade we have seen continuous migration of services from own infrastructure to cloud environment. Cloud services will continue to grow in future and give many benefits as it is flexible, scalable, and cost effective. In addition, it will create new application and services which may lead to creation of new job roles. To summarize, cloud computing will transform the ways we work and do businesses.

Popular Cloud Services

Amazon Web Service (https://aws.amazon.com)

Microsoft Azure (https://azure.microsoft.com)

Google Cloud Platform (https://cloud.google.com)

From the Author

This book is written to give you a basic understanding of the cloud computing services and business model. Hope I am successful in my endeavour and you are able to get the concept.

Your views and feedback may be communicated at the given email or message me on my social media pages.

Email address:

crossthebridgewithme@gmail.com

Instagram Id: Cross_the_bridge_with_me

LinkedIn URL:

https://www.linkedin.com/in/pankaj-gupta-0188135/

Related books

Telecom Technologies simplified for everyone:
Evolution of technologies and services

https://www.amazon.in/gp/product/B0B1MZ59MF

Understanding Customer Service: A must read for everyone in Service Industry

https://www.amazon.in/gp/product/B0BLT49RHB

www.ingramcontent.com/pod-product-compliance
Lightning Source LLC
Chambersburg PA
CBHW050315220526
45465CB00005B/2011

* 9 7 9 8 3 6 9 6 7 0 7 9 8 *